a Time for Hawking

Bruce Coughran

a Time for Hawking

Indra's Net Theater Press

A note about music: Original music created for this production is available to be licensed from Chris Houston at Implied Music.
Contact **chris@impliedmusic.com**, or **www.impliedmusic.com**

First Printing: June, 2022
ISBN: 978-1-7377171-0-2

Design by Ann Higgins
Printed in the United States of America

a Time for Hawking had its world premiere, presented by Indra's Net Theater (Bruce Coughran, Artistic Director; Alexandra Frappier, Managing Director) at Berkeley City Club in Berkeley, CA on Dec. 20, 2018, with the following cast:

STEPHEN HAWKING	Alan Coyne
JANE WILDE	Adrian Deane
JAYANT NARLIKAR	Tirumari Jothi

Director	Bruce Coughran
Sets	Sarah Phykitt
Lighting	Jim Cave
Costumes	Lisa Claybaugh
Original Music	Chris Houston
Choreography	Barbara Bernardo
Stage Manager	McKenna Moses

The play was nominated for four Bay Area Theatre Critics Circle Awards, and won the Award for Best Overall Production in the East Bay (Theaters 100 seats or less) for 2018.

CHARACTER	DESCRIPTION	AGE
Stephen Hawking	Cosmology Graduate Student	21
Jayant Narlikar	Cosmology Graduate Student	24
Jane Wilde	Spanish Literature Student	18

Time: Last hours before New Years, 1963
Place: Back yard of a home in St. Albans, on the outskirts of London

Notes:
1) Dennis Sciama is pronounced "Sharma"
2) Note that this is New Year's eve at the end of 1962, the beginning of 1963.

Acknowledgements:
The author would like to acknowledge the generous help of the many people who reviewed parts or all of the manuscript and offered suggestions. Most especially the very patient Paul Heller. There was also early and continuing help from John Feld and Ken McLeod, as well as from Judith Workman, John Bidwell, Gerry Oliva, Joel Feigin, Merula Furtado, Erma Wheatley, David Presti, Linda Hess, Naome Dragstedt, Alan Coyne, Adrian Deane and Heren Patel, and most especially, thanks for the constant support of my wonderful partner, Alexandra Frappier.

(STEPHEN HAWKING (at age 70) rolls onto the stage in an electric wheelchair. His body is withered, and his face is flattened against a support board on the side of his head. He rolls to center stage and stops. Computerized Voice (the "Stephen Hawking Voice") comes through the speakers, as he sits motionless.)

COMPUTERIZED VOICE

Can you hear me? My name is Stephen Hawking. I am a Cosmologist, which roughly means I study the Universe. It is a branch of Theoretical Physics that includes some very interesting questions. Like What happened at the beginning of the Universe and What is the nature of time. I feel very lucky to have chosen this field. Because of Einstein's Theory of General Relativity there seemed to be a real chance of making some progress, as opposed to other areas of Physics that did not have such a rich theory to build on. It did turn out that the five decades I have worked in this field were, indeed, such a time. We did make some great progress. But it also turned out that this was probably one of the few areas where I could have done any work at all, because it was purely conceptual in nature. In early 1963 I learned that my life might be very short, and hampered by a lack of control of my body. I was told I should expect no more than two years to live. I could not see how I could go on, with such a short time. But, fortunately for me, something happened while I was home for Christmas holidays that year. I attended a party for New Years Eve. Something happened there that I did not expect. And my life was forever changed because of it.

(The wheelchair drives to the back of the stage and stops. There is a light change as Stephen's face stops being crunched, and he slowly unravels his twisted body and becomes his 20 year-old, fully

functioning self. He steps forward just as JAYANT (24) stumbles in with two beer mugs. He gives one to Stephen as they continue their musings ...)

JAYANT

> But...

STEPHEN

> I know...it sounds odd ....but they had boys too. It was just the name.

JAYANT

> Well it sounds very strange to me.

STEPHEN

> I wouldn't have wanted to be anywhere else. It was a good place to grow up. A perfectly good place.

JAYANT

> In India, we wouldn't even think like that – wanting to be somewhere else. "Tat Twam Asi." The place you are, is the place you are. You don't think of the possibility of being someplace else. It just seems silly. "odd..."

STEPHEN

> Well, I guess so. Brits are big on that. Everyone secretly wants to be an Earl. I guess you would call it a form of envy. "what could have been..," "if only...."

JAYANT

> Well it seems very silly. You are where you are. Why spend time wanting to be somewhere else? And it is very nice here. Different from Cambridge.

STEPHEN

> Yes.

JAYANT

I was happy to meet your family. They seem delightful.

STEPHEN

There you go. I think most people would consider them a strange bunch.

JAYANT

Your sisters seem lovely.

STEPHEN

Huruff...

JAYANT

No, Stephen. They seem quite nice.

STEPHEN

Well, just don't get into a croquet match with them unless you have protective equipment. Hawking competition can be brutal ...and life threatening.

JAYANT

I think I would enjoy the competition.

STEPHEN

(snickers. Then)

What games do you play in India?

JAYANT

Oh..cricket, of course. My uncle taught me chess. My family was not much for sports though. Study was always more important.

STEPHEN

I can't imagine being the son of a mathematician. I had to learn it all myself. I guess I always thought I would be a doctor like my father. Until I discovered maths in Sixth Form at St. Albans. Saved me from Tropical Medicine anyway.

JAYANT

In our house it was the study that was considered valuable. Not so much what you were studying. Remember...my mother taught Sanskrit... so a Sanskrit poem was just as valuable as an equation....as long as you had a position to discuss.

STEPHEN

The Hawkings too. In that, we share a root. Nothing without a position. A spirited debate was our most cherished family possession.

JAYANT

Well...it seems to have served you well. The department is starting to notice you.

STEPHEN

Oh...I don't know. I'm not going to be co-authoring with Fred Hoyle anytime soon! I don't even have a thesis topic yet.

JAYANT

Don't worry. It will come. You have some good ideas. And you can calculate. It will come. An idea will come.

STEPHEN

I suppose. I'll find something to justify my fellowship.

(A beat.)

JAYANT

I think you would have liked India. You said you were only there a short time.

STEPHEN

A few weeks, really. My family was there for a year but I had to stay back to do my A-levels. My father didn't like Indian food, can you imagine that? They

rented a house in Luknow, where the governor of
Uttar Pradesh used to live...before being disgraced
for corruption and thrown out. And my father hired
the cook who had worked for the governor. And
had him cook ENGLISH FOOD! Can you imagine
that? I would have liked something more exciting.

JAYANT
(pointing)

That building looks very old. Is it a church?

STEPHEN

Yes. A very ugly one, at that. It is all that is left
of the ancient Roman city that used to be here.
Verulamium, it was called. Second biggest Roman
city in England. Next to London. The church was
named after Saint Alban, the first Christian martyr
in Britain, they say. A Roman centurion, executed
for his Christian beliefs. They named the town after
the church.

JAYANT

So, you have your teachers, too. Christian masters
where you grew up.

STEPHEN

It's hard for me to think of them that way...but I
guess.

*(Suddenly a girl enters. JANE (18) is carrying a glass of champagne
with not much left in it, giggling and stumbling a little.)*

JANE

Oh. Um....I'm sorry.

(beat)

I just wanted to come out and see the moon. I'll...
come back…

STEPHEN

        No...it's alright. It is a party after all.

(There is a brief awkward silence. Jayant leaps in...then wonders if he said the wrong thing.)

JAYANT

(to Jane)

        Happy New Year....or is it bad luck to say that before midnight?

JANE

        Oh....I don't know.

(a pause)

        I guess we should just declare it "not bad luck". Who wants there to be bad luck in the world, anyway? We'll just say it is not.

STEPHEN

        Right. Well. Not bad luck then.

JAYANT

(after a beat)

        Not bad luck then...

STEPHEN

        Alright. Well. My name is Stephen. This is Jayant.

JAYANT

        Hello.

JANE

(tries a little curtsey ....but then feels silly)

        Pleased to meet you. My name is Jane. Jane Wilde.

STEPHEN

(pointing at the glass)

> You seem to be enjoying the party.

JANE

(giggling a little, despite herself)

> Yes. A little.

(Another pause)

STEPHEN

> I grew up here. On Alma Road.

JANE

> Oh.

STEPHEN

> You are a friend of Diana's? ...or Basil's?...or...?

JANE

> Diana. Diana invited me. Are you...

STEPHEN

> I am a friend of Basil's. We were in school together.

(a beat)

> Well...it is a nice party...isn't it? Or perhaps you are
> bored? Which is why you have wandered out to the
> back porch to look at the moon.

JANE

> Oh...no. Not bored. It is a magnificent moon, isn't
> it? I love a full moon. There is something inspiring
> about it every time. Don't you think?

STEPHEN

> Surely. Surely enough to grab a look.

JANE

You don't like the moon? Doesn't the wonder of the heavens hold any interest for you? What are you, an accountant? Or perhaps a civil servant?

STEPHEN

No. Just a student. My civil service career went up in smoke, I'm afraid. When I missed my final exam.

JAYANT

He slept through it.

JANE

You didn't.

STEPHEN

I did not sleep ...exactly ...I just woke up a little late. I knew there was SOMETHING I had to do...but I just couldn't remember what it was. So I listened to the 'Ring Cycle' and didn't realize until the next day I had totally missed it. Luckily I got into University though. So I didn't need to make my way as an outcast.

JANE

You know... I know you.

STEPHEN

Do you?

JANE

Yes. I've seen you with Basil. And I know your sisters. They were with me in school. I think you were actually in my class for a couple of months. At the girls high school. Wasn't that you?

JAYANT

Stephen told me about that. He said he first went to the girl's school.

STEPHEN

I told him that it is just a name. They took boys before the age of ten.

JANE

And you were in my class for a few months.

STEPHEN

I don't remember. I was there. But...I'm afraid I don't....

JANE

Yes. I think I remember you. Your curly hair over your eyes...always hunched over your pencil.

STEPHEN

I was there for one term.

JANE

And I knew your sisters too, I saw them at school.

STEPHEN

They were younger, of course.

JANE

Yes. And your father was the beekeeper, right? He got a swarm of bees from our back garden once.

JAYANT

I didn't know your father was a farmer. I thought he was a doctor.

STEPHEN

Yes. He was. But he also kept bees.

JANE

He was the town's only beekeeper. We had to get him to come get the bees. When there was a swarm.

STEPHEN

Well. You certainly know about me.

JANE

Well...your family was so...

STEPHEN
(to Jayant)

See! They always thought we were odd here.

JANE

Well...you were! Diana told me she was invited to your house for dinner one night...and that everyone came to the table with something to read, and just ignored her the whole dinner! Even when she was the guest!

STEPHEN

Well...we don't like to waste that reading time.

JANE
(to Jayant)

I remember his mother...we would see her at the crossing each day, at the town traffic light. Waiting for your younger brother to come out of school...IN HER FUR COAT! I saw her every day.

(to Stephen)

So....you made it to University.

STEPHEN

Yes. I was at Oxford.

JANE

Oh yes. You were part of the storied "class of '59."

(to Jayant)

There were eleven of them going to Oxford

that year...a lot for our little school. The local paper made a big fuss about it. The "intellectual adventurers of our generation" they said. "Starting a whole new way of thinking"..."rejecting the old ways"...and all of that...

STEPHEN

Yes. Well...

JANE

(to Jayant)

We all thought it was a bit of tosh.

(a beat)

Well. Neither Oxford nor Cambridge were much interested in me. Much to the disappointment of my father. I am going to Westfield. Just finished my first term.

STEPHEN

I was surprised I was admitted to Oxford. I thought I messed up the entrance exam horribly. And everyone thought I was too young to even be applying.

JAYANT

Those entrance exams must have been very hard. Why did you take them early? Were you too advanced for your class?

STEPHEN

Hardly. I was never more than halfway up my class. My father had wanted me to go to Westminster (the Public School) instead of staying at Saint Albans school. But I was sick when the scholarship exam came up and never took it. I think I got just as good an education at Saint Albans. But I was pretty average as a student.

JANE

But hardly an average family!

(to Jayant)

We all considered the Hawkings to be pretty eccentric. They had a Gypsy Wagon they camped in for holiday!

STEPHEN

(shrugs)

There were two other boys from the year ahead of me who were going up for the Scholarship exam to Oxford. My dad always wanted me to go to University College, Oxford....because that's where he went...so I tagged along. I was shocked when I got the telegram saying I had gotten the scholarship.

JANE

So what did you study?

STEPHEN

Chemistry. I wanted to study maths but my father was dead set against it. "No future in it," he would say. And they didn't have a maths track at University College anyway. He wanted me to do biology and become a doctor. Chemistry was the compromise. I later switched to physics. Because it was an easier track.

JAYANT

Your "intellectual adventurer" likes to brag that he never did any work at Oxford. Says he never attended a lecture and left all his exams to the end.

STEPHEN

You could do that in physics.

JANE

> You didn't get looked down on?

STEPHEN

(laughing)

> At Oxford?? No. You were not supposed to be
> working hard at Oxford. It was considered very
> poor form. You were supposed to be so smart you
> didn't have to study ...or too good mannered to
> not try to appear to be. Certainly not to work hard.
> That was the lowest.

JAYANT

> Not at Cambridge.

STEPHEN

> No. I suppose not.

JANE

> You are at Cambridge now?

STEPHEN

> Yes. Jayant has the unfortunate burden of having
> me share his office.

JAYANT

> We are both Physics graduate students.

JANE

> Oh....well. It worked out then.

JAYANT

> Stephen likes to brag that he threatened them at
> Oxford. He says that he had such a bad reputation
> ...he was on the edge between the First Class
> degree and the Second Class. When they asked
> him what his plans were, he said if he got the First
> Class degree he would go to do graduate studies at
> Cambridge. If he got Second Class he would stay at
> Oxford.

21

STEPHEN

> They gave me the First Class.

(Laughter)

STEPHEN

(pointing to her empty glass)

> Ah....

JANE

> Oh....

(with mock politeness)

> That would be very kind of you, Sir.

STEPHEN

> I shall return, my children....

(Stephen takes her glass...and the mug from Jayant, who nods, and his own mug and leaves.)

JAYANT

> You have a very nice town here. Stephen was telling me it was once a great ancient city of the Romans.

JANE

> Oh, did he? Well, yes...I suppose it was. Now it is just a simple little town. You are from India.

JAYANT

> Yes...I grew up in Benaras. I was lucky enough to come here after my University degree to study with Doctor Fred Hoyle.

JANE

> The Television Physicist? I've seen him on the BBC!

JAYANT

> Yes.

JANE

Is Stephen studying under him as well?

JAYANT

No. He wanted to. But Dr. Hoyle was not taking on any additional graduate students. So he went to Dr. Sciama. Also a very good physicist.

JANE

But he doesn't have a show on the Tele.

JAYANT

No. He does not.

(a beat)

JANE

So. India. It sounds so exotic. Will you go back then? When you finish your studies? Or do you think you will stay in the West?

JAYANT

I do not know.

JANE

Oh.

JAYANT

India has great history. It had a thousand years of history ... literature and science... even BEFORE the Roman town was built here. An ancient history. It is a rich place. And it is home. But there is wonderful science being done here. There are so many more resources here. In the West. In England. At Cambridge, actually. It's hard to imagine leaving that. At least very soon. I hope I will go back someday. And maybe be able to take some Western science back with me.

JANE

Well...that is a grand ambition.

JAYANT

Of course we never know. We never know how much time we have, do we? I like to imagine that there will be time for it all. Like Narada.

JANE

Narada?

JAYANT

Narada. It's an old Sanskrit story. A famous ascetic named Narada was so devoted to his ascetic practices that the God Vishnu appeared to him and granted him any wish he desired. He asked to see the magical powers of Maya, the maker of the cosmic illusion.

So Vishnu said to Narada "come with me". As they walked they entered a huge desert, and presently Vishnu asked Narada to go to the next village and get him some water. When Narada came to the first house he knocked on the door and a very beautiful girl opened the door, and she so takes his breath away, that he forgets why he has come.

Being polite, she asks him in, and her father, seeing that he is a saint, welcomes him as a guest. After a while Narada falls deeply in love with the daughter, and marries her, and discovers the joys and hardships of peasant life.

After much time, the father dies, and Narada becomes the master of the lands. By this time he has three children, and soon a torrential rain begins that won't stop. He grabs his two older children in one hand and his smallest in his arms and tries to flee, calling to his wife to follow. He struggles through the deep, flowing water valiantly, but soon he is exhausted, and the waters are strong... and he slips, dropping his smallest child into the flowing water. As he tries to recover it he lets go of the other two children and they flow away in the flood, along with his wife, and he himself slips and hits his head on a rock and is carried away unconscious. When he comes to, stranded on a rock, he remembers his misfortunes and begins to weep uncontrollably. Suddenly he hears a familiar voice…"my child!" it exclaims. "Where is the water I asked you to bring me?" He looks out and sees the vast desert and Vishnu standing before him. "I have been waiting for almost half an hour!" Then Vishnu smiles, and says, "Now do you see the magical powers of Maya?"

JANE

(after thinking for a beat)

So...there is no time in imagination...or no limit to time...

JAYANT

Well. I don't think that....in Sanskrit philosophy .... Sorry, I am the son of a Sanskrit professor! But in Sanskrit philosophy there is not the idea of imagination...of one thing being imaginary and another thing being real. These stories are to show the real universe. The Vedas make no distinction.

JANE

Then the man, Narada, he actually lived that life? And then it was gone when Vishnu asked for the water? That is what we are supposed to believe?

JAYANT

Yes. Understanding the universe is difficult. In physics we pretend to make it simple...but it is still difficult.

JANE

(thinking)

There is a similar story ...by a South American poet, Borges. About a writer who lives in Prague, and

25

the Nazis come for him, because of his writing, and they take him to a bridge and prepare to shoot him. He worries about the story he has wanted to write for years and hasn't ever written. He looks at the rifles and can see the bullet in the barrel and then a bee flies in front of the gun and he watches as the bee goes by and it lights on a flower...and..... he starts to write. Day after day, hardly eating or sleeping, until after a year he finishes his story. His life's work. And he looks up... and the bullet hits him and he dies.

JAYANT

It is similar. He saw the limit of his time and lived his life anyway.

JANE

I think the story is supposed to show the richness of the world of the imagination.

JAYANT

But in India....we would not see a world of the imagination as a separate one. A separate world... from the one we know. That is the point. Maya is the one who makes us think that they are different. Vishnu sees that they are just the same.

JANE

What about physics? You are supposed to be a student of Western physics. Doesn't that...the... what are they...."laws of Physics"....don't they describe a world that is separate from what you imagine? A real world? Isn't that what you are after?

JAYANT

You'd be surprised. Modern physics...has some pretty strange aspects to it.

JANE

Really ...

JAYANT

And....most of physics...it only looks at a part
of the...physics...science, really ....takes just a
small part of the universe...something we can
understand...and tries to come up with things
we can know about it. That's what the laws of
physics are. What we can know. It's never...at least
according to Vishnu......it's never going to be the
whole story. The whole of reality.

JANE

Well wasn't I the misinformed one then. I thought
that in my physics class I was learning the answers.
That's what I liked about physics, actually. Learning
about the world...how the world worked. It was
just the maths that gave me trouble. Too many
formulas to learn.

JAYANT

You see. In India, we don't have the same language
as western physics. There are plenty of good
physicists, of course...but the language we are used
to...we describe things in different ways.

JANE

Such as....?

JAYANT

In Varanasi, where I grew up, there is a play that is
put on every year...the *Ramlila* it is called. It tells
the story of the god Ram...from the *Ramayana*
texts. It takes place over 31 days, with thousands
of actors, and tens of thousands come. The action
moves through cities and fields, lakes and the
banks of the great river Ganges. Even as a child, the
expanse of it...I knew you could never see it all in
one year — you would miss SOMETHING. You see

27

that it is the play of the Universe, with its richness and complexity... with fireworks at the end. After a while, the audience and actors seem to merge into one stream. The villages become the world of the gods...through the days..they all blur..and what you are left with is just the play of illusion.

JANE

That sounds spectacular.

JAYANT

So when you have that.... The West has this spectacular achievement — modern physics is an achievement. There is no reason that India cannot be.... cannot have both.

(Jayant looks away uncomfortably.)

JAYANT

I wonder if I should go find Stephen.

JANE

He seems like he can take care of himself.

(Jayant suddenly looks a little nervous.)

JAYANT

I was honored when Stephen offered to show me the town he grew up in. He is very bright....but being a new graduate student is not easy. The last few days he has seemed to be....perhaps I should go look for him.

JANE

I'm sure he is fine. So. Physics. You don't work on the bomb or something?

JAYANT

No. We are both in Cosmology.

JANE

>Oh. That sounds quite grand.

(Stephen re-enters with drinks and overhears this.)

STEPHEN

>Well...there are only two big areas in physics worth working on; Particle Physics and Cosmology. I thought that...well...most everyone is working in particle physics now...the world of the very small... it is fairly well defined. But Cosmology ....how the big things in the universe behave...well... I don't really suppose you are too much interested in this, are you?

JANE

>Sir! You cut me to the quick! Dare you think that a student of Spanish Poetry would not be interested in the workings of the universe? Is that what you are trying to study? The workings of the Universe?

JAYANT

>Yes...that is what it is.

(looking up)

>The movement of the stars and planets, and everything in the universe.

JANE

>Well...then that does sound grand. As long as you aren't trying to help Khrushchev and Kennedy blow us all to kingdom come.

(beat. Then, to Stephen)

>But I heard that you were a "ban the bomb" protester.

STEPHEN

>It needs to be done.

JANE

We came way too close.....it certainly seems.

STEPHEN

Yes we did.

JANE

Well...I have to admire you for it. I know everyone makes fun of you for going to those marches... but I think that....well....what did Field Marshal Montgomery say? He thinks there will be a nuclear war within a decade.

STEPHEN

It was stupid what Kennedy did in Cuba. Playing with nuclear war like that. We came way too close in Cuba. Thank God we didn't do it yet.

JANE

I just saw a lot of war memorials this summer when I was in Flanders. In the midst of all the rebuilding... memorials everywhere. My mother just says she'd rather have it all over in four minutes than see her husband and son go off to war for a third time.

(An awkward silence)

So, this Cosmology. How do you study the Cosmos? It's about looking at the stars?

JAYANT

We try to come up with theories ...about the stars. About the universe. Not just look at it.

STEPHEN

How the universe started. How it will end.

JANE

Theories...

JAYANT

Yes. And there are some marvelous theories. Quite beautiful ones.

JANE

I'm sure I couldn't see what a "beautiful theory" could be.

STEPHEN

Einstein's Theory of Relativity. That is beauty itself.

JANE

That's what made the bomb, didn't it?
E equals M C squared??

(Stephen and Jayant look at each other...do they really want to get into this??)

JAYANT

That was the Special Theory of Relativity. In Cosmology we deal mostly with the General Theory of Relativity.

STEPHEN

Gravity. Einstein's General Theory of Relativity has to do with Gravity. And it is beautiful.

JANE

Well..OK. You will have to prove that to me. I may just be a Westfield girl...a Spanish Lit major at that...but I took physics and you'll have to show me the beauty. I thought the theory of relativity was just that everything is relative. You can't know if you are moving or the thing you are looking at is moving so it is all relative. Right?

(Stephen and Jayant again look at each other. Jayant has heard this all before..and he feels the call of nature, so he bows and exits.)

JAYANT

Well...I think Stephen can tell you about that...I will

explore a little bit of the house...if you will excuse me.

(Despite themselves, being alone suddenly feels a little dangerous. Stephen steps back and then smiles a little lamely.)

STEPHEN

Hum...Alright. Here goes. It all starts with Newton. Newton's laws of motion. He came up with these equations for how things moved, forces and energy...

JANE

I know. F equals M A. I studied them. Had to learn them.

STEPHEN

And they worked really well. That is what our theories ultimately do...they predict how things will behave. And if they work, we call them "laws"....that is, if the predictions are born out in experiment...the theory works...until it doesn't. If the experiments don't follow the theory ...then the theory has to go. That's the only test. No matter how beautiful the theory ...if it doesn't match experiment, it has to go.

JANE

Of course.

STEPHEN

But you see...that's the thing. The theories aren't "real"...they just "work." That's really all we have.

JANE

Meaning??

STEPHEN

The laws that Newton came up with...they explained the motion of all the bodies in the

Universe, both the heavens and on earth...in ways we couldn't before that. And, as far as they were able to measure at the time, they worked. Predicted exactly.

JANE

So then what makes them beautiful?

STEPHEN

Well....we always...physicists always secretly, or not so secretly, want them to be simple and pure... and also to work. I guess that is the Greek ideal shining through. The best is when you come up with a theory that seems simple, inevitable, and surprising. After you come up with it, you see things you didn't see before...and it seems like it couldn't be any other way.

JANE

So Einstein's do this?

STEPHEN

So...you see...Newton came up with the laws... and he is revered because they worked so well. He came up with these simple equations to predict the movement of things...and then later Maxwell did the same thing with Electricity and Magnetism ....simple Laws that described how they would work...and they did.

JANE

Yes...I remember those formulas. I had to learn them too.

STEPHEN

But Newton had to come up with this force of Gravity. Brilliant insight really ...this force that drew all matter to each other, proportional to the mass and distance.

JANE

The apple falling on his head and all that.

STEPHEN

Yes...but...he never said what that force was...just said there was one. And the equations all worked so beautifully and it all made sense so we just accepted there was such a thing. Quite amazing he got away with it really.

JANE

Yes.

STEPHEN

But Maxwell...he had seen that electricity and magnetism ...that seemed different, were really two ways of looking at the same thing. And that there were forces "out there," but could be seen as two ways of looking at an electromagnetic "field," that existed in the universe. Einstein....made the bold move of daring to look at matter and energy that way.

JANE

E equals M C squared.

STEPHEN

Yes. And then. He went one step further.....he did the same thing with time and space.

JANE

Humm...what does that mean?

STEPHEN

So....Newton had said that this matter and the forces were moving in this empty "space." Which we think of quite easily. But, what Einstein did... and this is why it is beautiful...is he just said, if I treat it not as a three dimensional space that changes in time...but as a four-dimensional "space-time"...then the equations work. And explain Gravity as a kind of curvature of this four-dimensional space-time.

JANE

Well...that sounds terribly odd.

STEPHEN

Well it took him ten years to work out the maths... but it worked in the end. And you see...just like the electromagnetic field...the gravitational field (which looks, mathematically, very similar) is not something separate from space-time. It is an aspect of space-time itself. It makes the universe simple...in a funny kind of way. There doesn't need to be this separate "force"...of Gravity. And that, we call beautiful. You have to see that, don't you?

JANE

I suppose...I guess so. I don't know what a four-dimensional universe looks like.

STEPHEN

Well...no one does, really. The thing is that...soon after Einstein came up with this...they found experiments to test it.

JANE

How can you "test" such a thing?

STEPHEN

So, first of all, the equations, if you make a rough approximation, were the same as Newton's. So that was a good start.

There was also the fact that the theory explained a wobble that was known to exist in the orbit of Mercury, and hadn't been explained. Again, it just fell right out from the calculations. Something that had been unexplained for hundreds of years. After he finished the theory, he recalculated Mercury's orbit with the new equations and got the exact amount of wobble that had been observed.

Then there was the bending of light. So if gravity is a curvature of space, then light would have to travel through that same curved space. So you would see light bend when it went past a very massive object. That could be observed by looking at stars right next to the sun, and was, during a solar eclipse just a couple of years later. The amount of bending of the light was exactly as Einstein had predicted.

JANE

Hummm...

JANE

So, Einstein got evidence his theory was able to predict things. That doesn't seem like it necessarily makes it beautiful. It is just finding out how things are, right?

STEPHEN

Well. First of all, the equations that describe all this...the math that Einstein came up with after working on it for 10 years...is incredibly simple. One simple equation.

JANE

Like E equals M C squared?

STEPHEN

Yes...except it's G plus Lambda equals 8 Pi G over C to the fourth, times T. Not quite as tidy, or as easy to understand. But still incredibly simple.

JANE

Hummm

STEPHEN

But it works. No one really knows what it "means" in that sense ...just like particles. No one really knows what they "are".

(Jayant returns, with drinks.)

JAYANT

Oh...so we have made it to Elementary Particle Physics, I see.

JANE

Stephen was just telling me that no one really understands what things are...and physicists are fine with that...It's all relative. So you might as well just read poetry.

STEPHEN

That's not exactly what I said. People like particle physics because it is more well defined. They can smash atoms into each other and see if their particles behave like they think.

JAYANT

But they are no more "real" than gravity waves... less so really.

JANE

They are real. I studied them...atoms, protons, electrons... they are well known to be real. Right?

(Jayant and Stephen laugh again...)

STEPHEN

No more real than Don Quixote's windmills, I'm afraid.

(Jayant seems baffled by the reference. Jane notices.)

JANE

Don Quixote from a famous Spanish story. He fought legions of knights defending the honor of his lady, Dulcinea, but his servant just saw them as windmills. Quixote insisted they were knights.

STEPHEN

But Sancho, his servant, sees that they are just illusions.

JAYANT

We like to think of particles as little objects...like little balls...but that all went away in 1927.

STEPHEN

Quantum Mechanics? Have you heard of quantum mechanics??

JANE

Of course. I may be a Westfield girl but I'm not daft.

STEPHEN

Excuse me...I didn't mean...

JAYANT

In quantum mechanics...things don't behave as though they have a kind of reality we are used to thinking.

STEPHEN

The other great achievement of twentieth century physics. Did for the very small what Relativity did for the very large. And, it made modern science possible... even the bomb and everything...

JANE

God help us.

JAYANT

But it's compelling because looking at particles... you can come up with a really quite small number of particle "types" and build everything ...well... practically everything...from them. But all of particle physics...it's based on quantum mechanics...which is based on probability.

(At some point in here, Stephen finds a croquet set in the yard and they set up a croquet game and start playing.)

STEPHEN

So nothing is for certain...it just has a certain probability.

JAYANT

That certainty went away ...that was the Newtonian world...but that went away when Heisenberg and Bohr came up with quantum mechanics.

STEPHEN

There was a famous conference in 1927...everyone was there...Einstein, Bohr, Heisenberg, Planck, Madame Curie... everyone. When they came out, they knew that the deterministic universe, the universe Newton had described, was dead.

JAYANT

So Heisenberg came up with this idea. He said, it is impossible to know anything, really, about a particle when you are not observing it. So why not just say that it doesn't exist when you aren't looking at it. Then you make predictions on where it will be when you do look for it. And lo and behold...the maths all worked. You could absolutely predict the exact probability of where a particle might be when you look for it...with astounding accuracy ...but you had to give up the idea that it existed in the mean time, when you weren't looking.

STEPHEN

So you see...that leads to strange things like an electron being in one place...and then being in another place, with a certain probability, and you can predict these very accurately ....but the probability of it being in between is zero. So the electron has to jump from one place to another and never be in between...even for the smallest amount of time.

JAYANT

What they call the "quantum leap."

JANE

Well that sounds even stranger. Somehow I thought physics was about what was real....getting closer to what was real all the time.

STEPHEN

Well...that all went away ...as Jayant said...in 1927. Most physicists don't even talk about it anymore. "Philosophy "...they used to call it. Bohr, Heisenberg, Dirac, Schroedinger, that generation of physicists, they were all great philosophers. Now it is more or less just ignored...it doesn't matter — if you can find theories that work...that predict things...and that is what is important. Isn't it? Or at least that's all we have. What it means... the philosophy... The philosophy is just too daunting.

JANE

You might as well study poetry then.

STEPHEN

Well...perhaps.

(Jane "takes the stage.")

JANE

Y el mármol no hable lo que callan los hombres.
Lo esencial de la vida fenecida — la esperanza, el

dolar, el goce — siempre perdurará.
Ciegamente reclama duración el alma arbitraria
Cuando la tiene asegurada en vidas ajenas,
Cuando tú mismo eres el espejo y la réplica de
quienes no alcanzaron tu tiempo
Y otros seran (y son) tu inmoralidad en la tierra.

STEPHEN

No me suena a castellano ibérico... quizás más bien
del nuevo mundo. ¿Sudamericano?

JANE

Cono Sureño. ¡Muy bien! Excelente, Don Esteban.
¡Pero qué sorpresa!

(to Jayant)

Jorge Luis Borges. An Argentine poet. From the story.

STEPHEN

I spent the summer in Majorca when my father was
away on one of his many African trips. My mother
went to visit her friend in Majorca, who was married
to a poet. Robert Graves...so I learned a little
Spanish. And a little poetry.

JANE

Well...you <u>do</u> have surprises in you.

JAYANT

Well...I know nothing of Spain. So that was Spanish
then?

JANE

Yes. ...it is something like....

And Let not the marble say what men do not.
The essentials of a man's life — his hopes, his pain,
his delights — will abide forever.
But blindly the uncertain soul wants to continue,
When it is the lives of others that will make that
happen.

41

As you yourself are the mirror and image of those who did not live as long as you...
And others will be (and are) your immortality on earth.

....Something like that.

STEPHEN

It's about a gravestone...the marble...a gravestone.

JANE

Yes... "Inscription on Any Tomb" it is called. "Inscripción en cualquier sepulcro".

JAYANT

I do not think I understand it.

JANE

It is about life. What life is. And what it is not. It says that you live on....it is not what is written about you...written on the marble, but actual life that lives on. In others.

STEPHEN

It is about your work — your work lives on...but you do not.

JANE

No. I don't know what Mr. Graves taught you about poetry! If I may risk stating the obvious... First of all, it is not all in the words, of course. That would be prose. It IS the words....how they are chosen and how they sound, of course. But the essence of the poem is, I think, in the imagery ...the power, the insight...is in the way it is put together and opens up into something else. Here the imagery is all about life. What life is. By reducing it to some base of understanding...some simple understanding... we can miss the whole thing. That is what the poem says. It is a mistake to think you understand it by listening to the words. The poet wants us to listen to the words of course, but knows that you can't say what it means in words. That would be "the marble"...and it is the fool who tries to say "what the marble does not."

(She looks at Stephen. Stephen is at a bit of a loss for words.)

JAYANT

I suppose that it helps to know the poet.

JANE

Well it doesn't hurt. If you want to venture an opinion...but I hear the Hawkings are not shy about that, are they ?

STEPHEN

No.

JANE

The naive thing would be to believe the marble.

STEPHEN

Hummm...

JANE

(a deep breath)

Surely ... Why study poetry at all? It's easy to think it doesn't matter in any practical way. But what we are, what "this" is, we know it's not just.... We like to pretend that we can know. Put it in a book, some grand explanation. But ultimately we know it can't be that simple. Poetry — it makes that plain. We know we're never going to SAY exactly what the poem says.... It's what we learn while studying it. The insight that comes from looking. And appreciating the thing itself, of course. The beauty of the thing itself.

(to Jayant)

> I suppose you have poems in India too. The
> *Mahabharata*???

JAYANT

> The *Mahabharata* is 100,000 stanzas, 200,000
> lines. More than I can memorize.

STEPHEN

> A paltry excuse!

*(Stephen goes to the side and rummages through stuff on the edge
of the garden shack.)*

JAYANT

> It covers many great struggles and battles,
> between the gods and men, and ends with the
> beginning of our modern Yuga; the Kali Yuga.

JANE

> It is about the creation of the universe then....is it?

JAYANT

> In the Sanskrit view of time. The universe is created
> over and over again. The Yugas, there are four of
> them, each one of them is from 432,000 years to
> 1.8 Million years...the universe is destroyed at the
> end of each group of four Yugas, and a new one is
> created. The whole group of four cycles of Yugas
> is repeated 1,000 times to make a Kalpa, about
> 4.32 Million million years. And the Kalpas just keep
> going and going, in a cycle. So, you could say there
> was "a whole lot of creating going on."

JANE

(to Stephen)

> Is that what you would call Philosophy then??

STEPHEN

> No. That is what we call Cosmology!

44

JAYANT

In an Indian way...yes.

STEPHEN

So, you see. Our strange physics is not the only
thing complicated.

JANE

So is it complicated or simple? I thought you said
that the theories that were the most beautiful
were the ones that were simple.

STEPHEN

Well....

JANE

You don't really sound like you know what you are
after, at all.

STEPHEN

Well yes...

JANE

In poetry ...at least we know what we are trying
to do. In poetry... To experience poetry is to see
over and above reality. It is to discover that which
is beyond the physical, to experience another life
and another level of feeling. It is to wonder about
the world, to understand the nature of people, and,
most importantly, to be shared ...with another.
Poetry is about the inner life... the inner life
shared. It sounds like you don't know what you are
after....is it simplicity or complexity?

STEPHEN

(weakly)

Well...we are after ...something real. Something
that will last...at least people like Jayant here
who have actually found something to study. And

partnered with Fred Hoyle no less. I'm still trying to see what's worth doing. What's worth spending time on.

JAYANT

Some of our physics is simple… Particle physics for example…

STEPHEN

It is getting simpler…

JAYANT

It has a model.

STEPHEN

Yes… it does have a model.

(to Jane)

When I chose physics, there were only two fields that were really interesting…particle physics…and cosmology. Particle physics…despite the fact that it doesn't make sense…is really hot right now.

JAYANT

They have started to come up with a unified model for all the particles we know about…and all their associated fields

STEPHEN

each particle has a field associated with it

JAYANT

and they almost have it all tied up in a bow. They are searching for a theory that ties it all together. And they are close, so that is why everyone wants to work on it.

STEPHEN

The holy grail…a unified theory …it seems right around the corner.

JAYANT

The Standard Model...

STEPHEN

That's what they are calling it.

JANE

Seems a rather unimaginative name.

STEPHEN

Yes...not very poetic, is it?

JANE

So why aren't YOU searching for the grand unified theory. Where is your ambition?

STEPHEN

Well...for one thing most of the department at Cambridge don't think there is a unified theory ...one that comes from quantum mechanics anyway. And besides...to me, it is too much like Botany....classifying all these particles and their forces and...and everyone is working on it.

JAYANT

Whereas cosmology...well, we are considered an unrespectable bunch.

STEPHEN

Yeah. Tell her about Mr. X.

JANE

Mr. X? It sounds like a spy or something.

STEPHEN

Jayant just went to this conference on the Nature of Time. Always a problem, Time.

JAYANT

I went with Dr. Hoyle, and Dr. Sciama. Roger Penrose was there. Bondi, Gold. Hogarth, who organized the conference, presented a paper that was based on Feynman's PhD thesis.

STEPHEN

Richard Feynman. One of the best American Physicists. He cracked the outstanding problem of incorporating light into Quantum Theory. He was in the Manhattan project with Oppenheimer, and is now at Cal Tech. He used a totally new type of maths, called Path Integrals to do it… now it's being used for all kinds of things.

JAYANT

Well. The paper that started out the conference, presented by Hogarth, used the OTHER part of Feynman's thesis as the basis for explaining the Arrow of Time.

JANE

Alright.

STEPHEN

But it was something that Feynman had rejected …set aside really… something called the Wheeler-Feynman Absorber theory. And Hogarth was now applying it to cosmology.

JAYANT

So Feynman and Wheeler showed up at the conference. But when they found out that the conference was being recorded, Feynman demanded that they not record him…not even his name….he thought the whole thing was so flaky that he didn't want his name associated with it. So they agreed to call him Mr. X, in the proceedings.

STEPHEN

So you see… this is how little is thought of the state of cosmology these days.

JANE

> So why would you want to work on it?? Just because it is disreputable? That seems a little bit like spiting the horse by walking. You may get your way but you won't get anywhere.

STEPHEN

> But cosmology and gravity... it is all based on Einstein's general theory of Relativity... and that is...well...it is so brilliant that it seems like all the real work has been done...so hardly any of the really great minds are working on it. There seemed to be more of a place for me.

JANE

> Looking at stars?

STEPHEN

> Well...it's more than that. It is, ultimately, what brought the world into being in the first place.

JANE

> Creation? I thought you were an atheist ...a scientist.

JAYANT

> Oh no. A scientist is interested in creation, as well.

STEPHEN

> There is really no particular conflict between religion and science...except for things like transubstantiation, of course.

JANE

> So...you did go to Sunday School once! Or was this just another Hawking family debate subject?

STEPHEN

> I'm afraid I must confess. It was.

JANE

> Ah.

(again noticing Jayant is left out)

> Transubstantiation. It is a Christian doctrine. That the Eucharist, Communion, ...do you know what communion is?

JAYANT

> Yes. It is an offering you do in church, right? You get an offering of food from the Priests or something.

JANE

> Yes. Something like that. It is one of the holy sacraments. In all Christian churches really.

STEPHEN

> In the traditional Catholic interpretation, the "offering" ...the bread and wine the Priests give out... are actually transformed into the body and blood of Christ. That is called Transubstantiation. It is something the Church of England rejected. All of the so-called Protestant churches...a big break from the Catholic Church.

(Jayant makes a face.)

JANE

> Yes. It doesn't seem very scientific, does it?

JAYANT

> No.

STEPHEN

> But... other than things like that. Religion and Science are not incompatible. Heisenberg said that part of the confusion is that the two have different languages, and we confuse them. The heavens we talk about in religion are not the same

50

as the heavens into which we launch rockets and fly airplanes.

JANE

So… even you scientists acknowledge the place of religion.

STEPHEN

Well I don't know if I do. I am undecided. But Einstein, Heisenberg, Schroedinger, …they were all pretty religious. And they thought a lot about it. I think the general consensus for them is that modern physics is not incompatible with religion… it just doesn't have much to say about it. Quantum mechanics fully acknowledges that we don't know what matter is… much less anything like spirit. We always really knew that, Quantum mechanics just made it obvious.

JANE

You can't deny the place of spirit…even if you don't call it Religious, can you? I mean what would life be if it was just the material. It can't be just that. The world of the mind… Art, music, beauty… it can't be just your intellectual "beauty" of your equations. That is kind of obvious, isn't it? Poetry would not exist in that world. Or music.

JAYANT

For Hindus…the world is just the play of the Gods. We don't see any conflict between science and religion.

STEPHEN

Yes…you don't have the whole "world created in seven days" problem.

JAYANT

Vishnu and Shiva are constantly creating and destroying the universe.

JANE

Is that what you are hanging this on? Creation?? I
thought you scientists didn't have to worry about
creation… I thought that was what you left to
religion.

STEPHEN

No. It is actually one of the big questions in
Cosmology. Up until Einstein, everyone accepted
that the universe was pretty much as it always had
been. Changing… yes… but more or less uniform
and similar.

JAYANT

But under General Relativity… you can adjust the
constants… that means that under certain ways of
looking at it, the universe is always expanding.

STEPHEN

And there is evidence for this… more evidence all
the time…

JAYANT

But that means not only that the universe is
arbitrary… but also that it could have, at one point,
not existed at all.

JANE

You'll have to explain that to me.

JAYANT

There is a theory that the universe is expanding…
and that it will keep expanding… but that also
means you can trace it back in time… and see
where it started… where the expansion started.
And that would have had to have happened really
fast, all at once, the universe shot out. Bang.

STEPHEN

Hoyle… Jayant's wonderful boss, on his BBC
show, called it "The Big Bang"… it's a name that is
starting to stick.

JAYANT

> He said it, of course, to show how ridiculous it is! Not only does it mean that everything could be different...if you had a different constant in the equation...Einstein's equation... but also that there is a literal creation at some given, calculable time. Something most physicists would be very reluctant to accept.

JANE

> So....the physicists have to come to the question of....dare I say it?....a God?

JAYANT

> Well...I know people like to think that...but no. It does not require that...but still, Hoyle's view of a steady state universe...one that always follows the laws of physics...and those laws don't have to change at some point in time...as would have to happen with a big bang... You see at the moment of the big bang, or really close to it, the laws of physics break down. You couldn't have time, space, motion like we think of them...there would have to be something else. So the steady state universe... just seems to be more....makes more sense....and is more beautiful.

JANE

> There is that word beautiful again.

STEPHEN

> Well...science should be beautiful...mathematics, physics...that is one way you know it is true. It just seems like there should be a simple, "beautiful" way to explain how the universe works.

JANE

> Why? Because God is beautiful? You seem to be making the case for a God.

STEPHEN

No. Because it should be. Newton's Laws were simple and they worked. It was so stunning... because they worked... and explained so much. That is beautiful. We follow the math... work on these theories... in search of where it all comes together... where it all works... and is... hopefully... simple and beautiful.

JAYANT

And "the big bang" is NOT beautiful...it simply makes no sense...even Einstein saw that.

JANE

But didn't you say that Quantum Mechanics made no sense either?

(Both men laugh)

STEPHEN

Yes... but they don't make sense in different ways... and they don't work together... so I guess maybe that makes it alright.

JANE

For future "scientists" you seem to have a funny view of what is important... truth-wise.

(Pause)

So... are you really saying that this physics that is supposed to be so beautiful... is based on two theories... that are both based on things that are clearly absurd?

(The boys look at each other.)

STEPHEN

Yes.

JAYANT

But it's worse than that.

STEPHEN

    Because the two theories...quantum mechanics and general relativity...

JAYANT

    They contradict each other.

STEPHEN

    At the mathematical level...the equations...they can't both be true. They contradict each other.

JANE

    So...does that mean they are both not true?

STEPHEN AND JAYANT (together)
    NO!

JAYANT

    You see...Miss Jane. They both work. Incredibly well. They are both amazingly accurate at predicting things. Just different things.

STEPHEN

    So we can't throw them out. We keep using them.

JANE

    I can't believe this.

JAYANT

    Because they are so useful. It's just on the theoretical basis that they contradict each other.

STEPHEN

    So we live with it.

JANE
(laughing)

    And I thought...somehow...that science was supposed to be logical...more "real".

(a beat)

> In poetry, we look for truth. I don't know how
> that relates to reality...it is just "true". You look
> for something that has a deep and profound
> resonance that says something about the subject
> of the poem that can't be said any other way. All
> those truths can co-exist, I suppose you could say.
> There are different points of view. But when you
> get deep enough....you couldn't say that poems, or
> stories, "contradict" each other. That just does not
> sound right.

STEPHEN

> So it is with the world of 20th century physics. We
> don't know how to make sense of it...we just try to
> find more clever ways to make the equations make
> sense with each other.

JANE

> So this is what the intellectual adventurers of our
> generation are doing???

STEPHEN
(psuedo-defensively)

> Yes. When we are not drinking beer or playing
> croquet.

JAYANT

(to Stephen)

> Or listening to that "Wagner".

JANE

> Oh. You like opera?

STEPHEN

> I have the ring cycle on a big tape in my room...the
> whole thing. I do play it way too much.

JAYANT

> I think it is the reason Stephen keeps buying me beer. As a fellow at Kings College, I have access to the Library. They have a very good record collection. He gets me to check them out.

STEPHEN

> And I have a very good tape recorder.

JANE

> I love music too. I adore Brahms.

STEPHEN

> Ugh. Brahms. He was a very poor orchestrator you know.

JANE

(a little taken aback)

> I find it quite beautiful.

STEPHEN

> Pish. Brahms...no. As worthless as ballet music.

JANE

> Ballet Music!? Now you surely must be joking. I love the ballet.

STEPHEN

> Compared to Opera...it is so much more trivial. Ballet music is not worth listening to.

JANE

> I just don't know how you can say that.

(a beat.)

> The beauty of the ballet comes through the music. I may not have much of a musical education, but I have been going to the ballet since childhood, and I know the beauty in there. I feel it. It can't be all

about Opera. How can you say that? Piano music?
Beethoven? Rachmaninoff?

STEPHEN

Rachmaninoff? Only worthy of the musical
dustbin.

JANE

Ahah!

*(Jane is aghast. She also knows that Stephen might be just trying
to get a rise out of her. She takes the croquet mallet out of his hand
and goes over to hit a ball. It goes through the wicket. She hands the
mallet to Jayant who feels caught in the middle.)*

A long beat.

JANE

So.... this Ring...what is it? It's a Wagner Opera,
right? Tell me. What makes it so "elevated among
musics?"

STEPHEN

Well...it's not just an Opera. For one thing...
it is four Operas... meant to be performed on
successive nights. 16 or 20 hours of music and
drama, depending on how fast you play it. It
is so vast... in its scope and ambition. And it
doesn't follow any pre-existing rules of Opera.
It is always new and still all part of a whole. And
the story... it starts in the Rhine... I mean in
the middle of the Rhine. It starts from nothing.
The faintest rumblings of some unrecognizable
horn. And slowly the music seeps in and creates
the world of the opera. The three water nymphs
are guarding this magical gold under the river
that is the substance of the world. This dwarf
comes along... and, well, he is an ugly dwarf,
and they are beautiful water nymphs and they
abuse him terribly... toying with his lust and then

cruelly mocking him. He is so devastated that he denounces love… that gives him the power to steal the gold, which he does, and makes this ring (with all the power of the universe) which is what they fight over for the rest of the opera… through generations… and battles among the Gods and men… And in the end the Gods muck it up terribly and destroy the whole world. Twenty hours of opera later… But it's the music. Such music as you can't believe. It is grand and profound and big and reaches to the heights and the depths…

(beat)

You must hear it for yourself.

JANE

Hum. Well… perhaps I will one day.

JAYANT

He plays it a lot. I don't know how he gets any physics done.

JANE

So do you do experiments or what? What does a cosmology graduate student actually do?

JAYANT

No. We are theoretical physicists.

STEPHEN

I tried to do some experimental work. It was fun but I wasn't as good at it as I am at theoretical work. So we both do theoretical work.

JANE

Which means what? That you just think about things? Come up with concepts? Theories? That sounds boring.

JAYANT

Not at all! It is very exciting.

STEPHEN

The truth is it is all in the maths. We try to come up with equations that make other equations make sense...and hence solve the underlying truths of the universe.

JANE

You mean the theories that don't make sense that contradict each other in the end...but the math works so you are happy.

STEPHEN

We try to be happy.

JANE

Hunh....

JAYANT

We try to solve equations...yes...that is what we do. Solve equations and come up with new ones that connect the other equations. Stephen is very good at visualizing them...he is quite good at putting it all together.

STEPHEN

But it is all, for us anyway, in our heads and on the blackboard.

JAYANT

So for our group, that is what we do. It will be others who do experiments....

STEPHEN

...and build bombs and things. We will live in the world of the mind.

JANE

I'm sure it is all terribly important and prestigious and all that. But it sounds like a waste of time. If it's all theoretical and doesn't make sense at the end.

JAYANT

But that is why it is exciting! We try to solve the questions that seem unanswerable....to find new ways around the apparent contradictions.

STEPHEN

And we have some fun...on occasion.

JAYANT

The first seminar Stephen came to, when he was a new Graduate Student, Dr. Hoyle was trying to counter the evidence for this "big bang".

STEPHEN

....so Jayant and Hoyle came up with this paper that explains the arrow of time in an expanding universe...and kept the steady state model alive.

JAYANT

....the arrow of time...you see...in all of the equations of physics...time doesn't have to travel forward. So the question is always why does it seem to? Why do we remember the past and not the future? We can knock the tea cup off the table and it shatters and the tea goes everywhere...but the tea never comes back together and the teacup reassembles. Why ?

STEPHEN

...anyway...they had this paper

JAYANT

And Stephen went to the lecture....where Doctor Hoyle presented it. During the question period he told Hoyle that the theory didn't hold up because in a steady state universe, the masses in the equations would become infinite.

STEPHEN

He asked how I knew that. I said because I had
done the calculations.

JAYANT

Everyone thought that he meant he had just
calculated it in his head...no one else knew that I
had given him a copy of the paper the week before
and he had done all the calculations on his own and
found the flaw.

STEPHEN

I suppose it didn't hurt my reputation...

JANE

Well...you do have a bit of cheek about you after all.

JAYANT

Dr. Hoyle was furious.

JANE

What do you mean the "arrow of time" doesn't
need to go forward? Time is moving. Things
happen. That seems the most obvious thing.

STEPHEN

Time seems to go forward. Doesn't it? But it
doesn't have to, according to the laws of physics.
The equations work the same both ways. If time is
positive, or negative.

JANE
(skeptical)

Yes...

STEPHEN

So...stand over there.   I'll stand here.

JANE

Alright.

STEPHEN

If you move toward me...go ahead do it.

(they take locations at either side of the stage. She walks toward him.)

If you just look at me...not around...you have the experience of us coming together.

JANE

(with a little giggle)

Yes...

(As they get face-to-face they pause and smile at each other.)

STEPHEN

Alright. Now go back.

(she does)

Now I'll walk toward you....just looking at me...not around.

(he does. When they get face-to-face)

The experience is the same, right? Except for the fact that you experience the "walking"...but let's forget that for the moment. If it weren't for that...we wouldn't really know who was moving and who was staying still. Unless you look outside at the yard or the view or at Jayant (ick!) That...is classical relativity. You can't tell who is standing still or who is moving unless you say moving "relative to some outside frame of reference". So if we are in space ships, neither of us would know who was moving and who was not.

JANE

Alright.

STEPHEN

So now.

(Jayant rolls a croquet ball to Stephen. He knows what Stephen is doing.)

STEPHEN

> Let's say I have the coordination to roll this croquet ball at a constant speed, let's forget that I can't even come close...but let's say it is always exactly the same.

> If I am rolling the ball at you, standing here, it is going to take more time to get to you than if I roll the ball while I am walking toward you. It will come faster.

(He does. She catches the ball when it rolls to her. When they get face-to-face, they smile and then he backs up to the place he started.)

> But if you roll the ball while I am moving backwards...

(She rolls the ball to Stephen as Stephen walks backward.)

> Then it will take more time...and seem to go slower. Jayant sees what is going on...because he is an outside observer. He is our frame of reference.

JAYANT

> How flattering.

(Stephen takes her hand and they both move across the floor. As they do he rolls the ball but it stays between them as they are walking at about the same speed, as he says...)

STEPHEN

> But if we were both moving...on a moving pavement, let's say, then we would zip by old Jayant and the ball would seem to be going much faster to him....even though everything would look the same to us. So the speed...how FAR the ball goes in a certain TIME. Would look different. It

would seem to be much faster to Jayant because it
went so much farther in the same amount of time.

JANE

I'm waiting for the punch-line.

STEPHEN

So...what Einstein's first theory of relativity says...is
that if the balls are pieces of light...that the speed
of light looks the same to all three of us. The speed
of light is a constant to all three observers.

JANE

That doesn't make sense. How can that be? Why
does the speed of light need to be constant?

STEPHEN

The easiest answer is that the math works out that
way. But it was actually observed to be true, even
before Einstein. An experiment called Michelson-
Morley. They looked for the change in the speed
of light going with the spinning earth and against
it...and they couldn't find it. So Einstein just
incorporated it as a principle in his theory, and it
opened everything up. The maths worked.

JANE

Well... You already told me that Physics doesn't
really make sense. So my stance in Medieval
Spanish Poetry seems to have been well defended
against the intellectual adventurers of my
generation.

JAYANT

There are more things in heaven and earth than are
dreamt of in your philosophy ...

STEPHEN

But do you see what that means? If the speed of
light is always the same to all observers ....and
Jayant clearly sees the ball go farther (on the

moving pavement) than we do...then to keep the speed (distance vs. time) the same...the TIME has to change.

JAYANT

This is why they say that Relativity means time slows down on a spaceship ...because the speed of light stays the same....Einstein essentially said that it is not time and space that are constant...it is the speed of light that is constant ...and he was right.... so there we are.

JANE

(pauses)

Are you really saying that time...time itself...slows down or speeds up?

STEPHEN

A surprisingly difficult question to answer. But... essentially, yes. In relativity ...it means that if you have a twin that is traveling really fast....or even if you think of a twin higher up on the globe...

JAYANT

Because the earth is spinning so someone higher up ...on a mountain in the Himalaya, for instance, will be moving faster than someone in a low place

STEPHEN

Like Saint Albans...the two twins will not age the same. When they get together...to celebrate their birthday... they won't be exactly the same age.

JAYANT

The one who was on the mountaintop would be just a little tiny bit younger.

JANE

That just doesn't sound right.

STEPHEN

Yeah....but THAT is relativity ...the way it works.

JANE

In theory ....

STEPHEN

Oh no. It has actually been demonstrated ...pretty conclusively ...over and over again.

JAYANT

The Gravitational Red-Shift.

STEPHEN

Remember Einstein figured this out fifty years ago. Everyone tested it. It's now pretty clear that this is how it is.

JAYANT

A lot of experiments.

STEPHEN

To back up the equations ...which are pretty well accepted by everyone.

JANE

So you are telling me that time....the way time goes...flows....is not the same for everyone.

STEPHEN

Strikingly, yes...that is what it means.

JAYANT

It was the end of the idea of objective time. Everyone's time is necessarily different and individual.

STEPHEN

They said it was the end of simultaneity. You can't say that two things happen "at the same time." Time is local. Subjective.

JAYANT

> And in particle physics...this vast theory that they are about to tie up with a bow, the Standard Model, time does go backwards sometimes. Certain particle functions are described with time going backwards.

STEPHEN

> Feynman came up with these diagrams — brilliant — that show all the paths a particle can take... let you actually visualize this craziness. A huge breakthough....you see them all the time now. One of those axes is time.

JAYANT

> And some of those paths in the diagrams are going backwards in time. You can even think of something like a positron as an electron going backwards in time. It has become very useful to think of things this way.

STEPHEN

> But you have to accept that time is sometimes going backwards.

JANE

(A beat for this to sink in.)

> So there is this story by Borges....who wrote the poem. I was telling it to Jayant...about a writer...captured by a Nazi SS officer, about to be executed...and he thinks of his unfinished story ...and he sees a bee and watches as the bee stops, as if it were about to light on a flower...and he starts to write...and writes, page after page, for a year. He writes the last sentence ....looks at this finished story ...his life's work...and looks up and sees the bullet coming toward him, and he dies.

STEPHEN

(a beat)

You see, time is a problem. THE problem really. Time seems to flow...we all experience it...from the present to the past..or the present to the future, depending on how you think of it.

JAYANT

But the big thing about general relativity...is that time is not a separate thing...like it is in special relativity. The way we think about a 3-dimentional world, where things happen in time..in order....one after another.

STEPHEN

But in general relativity...which, like I said, seems to be shown over and over again to describe the real world as it is...time-space is one four-dimensional thing. So time and space can twist and stretch and do all kinds of funny things...maybe even like the story...but it should, just like space...it shouldn't be disappearing...or passing...it should be malleable ...like space. So why does it seem to "flow"?

JAYANT

And would it flow if there wasn't an observer to watch it flow? If we weren't here...would there still be time? Are we creating it?

STEPHEN

Or is time just our experience of it? The philosophers, Hume, Kant and them, they thought that time was only a human creation. Something we use to make sense of our experience.

JAYANT

The math doesn't have anything like the flow of time...it just looks like a four-dimensional "thing" ....without flowing time. But we all seem to see flowing time. In three dimensions, you know that a place... "here"...is only defined by the observer. Your "here" is different from my "here". But we can't think of time that way, at least we don't. We don't think of "now" that way...my "now" is different from your "now". But that is the way it looks.

STEPHEN

So that generation of scientists...after relativity... pretty much accepted that the flow of time was only a creation of the observer. Einstein, famously sent a note to the wife of a friend who had died... "We men of science know... that time, is only a very persistent illusion". He was talking about the flow of time...as an illusion.

JAYANT

It is one of the great mysteries of twentieth century physics...Cosmology ...is partly about how the world started...if it came from a big bang, time had to START there...and space too.

STEPHEN

You can see why physicists are a confused bunch.

JANE

Even though they put on a good front....

(she holds her hand to her forehead)

Oh...I think I have had too much champagne....

STEPHEN

Nonsense....I don't think there is such a thing.

JAYANT

(getting up and taking her empty glass)

And besides...it is not midnight yet....don't you have to drink something special at midnight? A toast, yes?

STEPHEN

        A toast!

JAYANT

        Yes...a toast...not a piece of bread, right?

STEPHEN AND JANE

(laughing)

        No....

STEPHEN

        It is the same word...but it means a special drink...
        together...to celebrate...or hope for the future...

JAYANT

        Yes...a toast then. So you need a full glass for that.
        I'll be right back.

(After he is gone there is a prolonged awkward silence. Stephen picks up a croquet mallet and tries to start playing again ...but the energy is not there. He takes a few swipes...when he goes over to retrieve the croquet ball he stumbles awkwardly ...maybe out of Jane's sight.... She notices, but does not react. He puts the croquet mallet down and comes back toward Jane. He smiles uncomfortably and sits. There is a long pause.)

JANE

        So....you'll spend your life at this? Studying the
        Universe?

STEPHEN

        What else is there?

(beat)

        Oh...I guess there is Opera....

JANE

        Oh...yes. The Ring Cycle.

STEPHEN

Yes.

(Another awkward pause)

JANE

Well, I think it's marvelous ...a wonderful way to spend your time...twisted or slowed up or whatever it may be...

STEPHEN

Yes

JANE

Worthy of one of our...Saint Albans' Intellectual Adventurers of 1959!

STEPHEN

If the world doesn't blow itself up ...

JANE

Yes.

(a beat, looking up)

...and we'll go to the moon! If you believe the Americans and the Russians...

STEPHEN

It will be fascinating when it all comes together... that's what everyone hopes for...some grand theory that brings everything together.

JANE

It sounds exciting ...I guess it's just equations and such...but it does sound grand.

(pause)

Even the word... "Cosmology." It does sound grand, doesn't it?

(Stephen smiles weakly. Another pause.)

JANE

And you'll be there…I bet…when they figure it all out. I bet you will.

(Stephen is strangely distant. Jane can't figure out what to make of it.)

STEPHEN

Oh…I don't know…

JANE

Our…our intellectual adventurer…

(Stephen says nothing.)

JANE

I'll just have Cervantes and Borges to….

STEPHEN

(hesitantly)

Jane…it was very nice to meet you. I'm glad I had a chance to meet you.

JANE

(a little flustered)

Well….yes…for me too. I guess I feel I…well…we….I guess we haven't ever talked…but…

STEPHEN

Oh…yeah…St Albans School for Girls and all that.

(Jane doesn't know what to say.)

I like you, Jane.

(as soon as he says this he is immediately awkward. Jane smiles, and he is just more uncomfortable)

JANE

> Well...I...

STEPHEN

> I'm sorry ...I didn't mean to...to say that.

JANE

> No...I mean...it's OK.....I.....

(There is a big pause).

STEPHEN

> I....I've gotten kind of clumsy. The last couple of terms at Oxford I was just running into things... dropping things...I didn't think much about it.

JANE

> Oh...

STEPHEN

> Then I had a couple of falls...

(beat)

> At first they just told me to drink less beer.

(beat)

> So when I was back during the summer I saw my doctor and he finally had me go in for some tests. It is just...well...I might not really have...to be the

(Jane says nothing...she just waits for him.)

> Well...you see...the tests...they say I have this bad disease. A very rare disease. In my nervous system. It's bad.

JANE

> Is it... M.S.?

STEPHEN

> No...it's not M.S. But it is degenerative. I'm going to just get worse. There is no cure.

JANE

> Stephen...I am so sorry.

STEPHEN

> I'm sorry to....I mean...I don't really even know you ...really. It's just...I thought I should...I don't know...I thought I should tell you.

JANE

> Well.

(a beat)

STEPHEN

> It's called Motor Neuron Disease ...Lou Gehrig, the baseball player, had it. They said I was an abnormal case...usually people older get it. But it is not M.S. But it is....I'll just start to lose the use of my muscles. One by one. Inevitably. Yes. Not such a cheery prospect.

JANE

> Oh Stephen.

STEPHEN

> Yeah... you see...I probably won't be around very long. There is a good chance I won't even be here long enough to finish my thesis. So....I don't think I will be solving the mysteries of the universe. I probably won't even make it out of school.

(Jayant returns with three glasses of champagne. He doesn't quite notice what has happened between Stephen and Jane.)

JAYANT

> I like your western New Year's Eve parties...this is not like our new year's in India...of course we have

lots of festivals...Diwali. I guess that's our New Years. It is the start of the Hindu calendar. And we do celebrate. I'll always remember the first Diwalis when I was young. The Festival of Lights. Candles on every doorstep. So many sweets. Fireworks. In people's hands, sparks flying in every direction... they seemed like magic.

I imagined there was all kinds of magic left in the world to find. There will be fireworks, yes?

STEPHEN

Yes.

JAYANT

So, have you solved the problem of expanding and collapsing time??

JANE

No. I don't think we quite sorted that one yet.

JAYANT

Well, good. That is what we need as graduate students...a problem to solve. Without that we are nothing. Did Stephen tell you his idea of how to study time??

JANE

No. I don't think he did.

JAYANT

Well...he wants to look at these funny things called Black Holes.

JANE

What?

JAYANT

I know...you can't really look at a Black Hole...right? It's black! And besides, I don't know that we will ever really find one....no one has seen one and no

one may ever. Because they give off no light…so
how could you see them??

JANE

Well what are they?

STEPHEN

(trying to put up a good front)

In 1939 Robert Oppenheimer wrote a paper that
followed the equations of Einstein's theories and
he was able to show that under certain conditions
…ones that should be pretty common actually …a
star could collapse and the gravity would get more
intense as it imploded…and it would just keep
going…until you had something that is so dense
and had such great gravity that even light could
not escape. It would be black…because the light
could not escape. Strange, eh?

JANE

So, they are just theoretical?

STEPHEN

For the most part. We'll probably never see them.

JANE

But you could…I mean…you might be able to,
someday?

STEPHEN

Maybe…but not within our lifetimes.

JAYANT

It's hard to figure out a way to see them. Seems
doomed. But maybe someone will figure out a
clever way.

JANE

Wouldn't you count on that? Someone will be
clever enough to figure out the problem. To see

what you don't think it is possible to see. Don't you think?

STEPHEN

Maybe someday ...

JANE

But that day could be soon. And then there would be new possibilities ...that you couldn't see or conceive of when you think it is all doomed. Right?

STEPHEN

I suppose.

JANE

And time might stretch and bend in some unexpected way...and open up in a way you never expected...

STEPHEN

Perhaps.... And perhaps they would just be chasing windmills....that they pretended were battling knights.

JANE

To defend the honor of the lady...

JAYANT

From the story ....

STEPHEN

Or maybe they were really looking at a bullet that had just been fired....

JANE

Or a bee...that would lead to completing their life's work...Or a whole beautiful, full life with children and ...happiness....and...

(Jayant suddenly doesn't know what is happening. Something is

clearly going on between Stephen and Jane that he is not a part of. He hesitates uncomfortably.)

JAYANT

>Yeah....well. You see. The problem with time...
>In physics time is just...in a four-dimensional
>Space-Time... You and Me and Stephen aren't
>experiencing this moment together...well, we are...
>but in Relativity...we're not.

STEPHEN

(weakly ...his heart is really not in this any more...but trying to help out Jayant who is still going)

>You see Einstein discovered something terribly
>disturbing. This beautiful theory explained so
>much, explained Gravity, which had been a total
>mystery before that. But the theory predicted
>that...just like with Newtonian gravity....that
>because everything was attracted to each other...
>the Universe...that is, all of the matter in the
>universe would pull together and create a big
>clump. Inevitably. No cure.

JAYANT

>And Einstein, like pretty much everyone else at
>that point, believed that the universe, the Cosmos,
>was steady state. But the theory didn't allow for
>that.

STEPHEN

>Inevitable destruction.

JAYANT

>So everyone ...everyone in cosmology that is...
>is trying to find a way to make the Universe last.
>The Steady State Model...we talked about...
>What makes it work? It is the exciting work...and
>Cambridge is the place.

STEPHEN

> Hoyle is at the heart of it. And Jayant. They are the ones trying to save the universe.

JANE

> So Einstein found out that his famous equations were wrong??

STEPHEN

> Well...no. He fudged. He created a fudge factor.

JAYANT

> The cosmological constant.

STEPHEN

> Yes. But it was just a fudge factor. It introduced a small force that was enough to keep the universe from collapsing. But no one believes it anymore. Even Einstein called it a mistake. Because....we are making measurements now...by looking at stars in a certain way....and it looks like the universe is not static at all. It is expanding.

JAYANT

> We are coming up with ways. That paper that Stephen embarrassed Dr. Hoyle with...at the lecture where he said he had done the math....that was an attempt. We still are making progress. We'll get there.

JANE

> What a strange journey you two are on.

STEPHEN

> You are on it too. We just do the maths to try to describe it. The Universe just is.

(Sound of fireworks starts in the distance.)

JAYANT

    I must go see.

(looking back at Stephen and Jane as he starts to exit.)

    There is a better view from the garden.

STEPHEN

    How about you go. We'll come in a minute.

JANE

    Yes. We'll come in a bit. You go...it's quite a
    sight, the fireworks here. For a small town on the
    outskirts of London anyway. Please. Go.

*(Jayant hesitates ...but then goes. Jane and Stephen are alone
together. Clearly a bond had grown between them. Jane glances
over to where the fireworks are. But then looks back at Stephen. She
walks over and sits beside him.)*

JANE

    So it doesn't sound like you know so much about
    time after all. You don't even know how to put it in
    your equations.

STEPHEN

    Well. One knows when there is not enough of it left
    to do anything.

*(Another awkward silence. The Sound of fireworks rises. A song starts
playing...from the party. Maybe a waltz. Jane looks tenderly at
Stephen who is obviously uncomfortable. She stands up.)*

JANE

    Dance with me.

STEPHEN

    Oh. I don't know...I don't...

JANE

It's New Years Eve....It's Midnight. I think there is a universal obligation to dance when there is music at such a time. Isn't there? There must be a law. A law of nature. I'm sure there is.

(a beat)

Dance with me.

(Stephen shakes his head weakly. Jane stands defiantly. She curtsies quite properly. She holds out her hand to him.)

JANE

Mister Stephen Hawking. Dance with me.

(Something in her insistence overcomes his gloom. He stands and bows to her formally.)

(They dance.)

(At the end they are close. Stephen might be suppressing tears.)

JANE

I think...that even your physics says that time is what we make of it. It only exists from the point of view of the observer.

STEPHEN

You WERE paying attention.

JANE

And, it only flows based on what we see. Right?

STEPHEN

Yes. As far as we know.

JANE

I'll tell you what I know, Stephen Hawking. I know you have a lot to give. Your time is not up ...not yet.

And with what time you have...you will be able to do some great things, I bet.

(Jane pulls back. Takes her sweater and puts it on. She turns and looks right at him.)

JANE

So. Call me. Let's go out on a date. Call me soon.

STEPHEN

Alright. I will.

JANE

Make sure that you do.

STEPHEN

It's been a wonderful time.

JANE

Yes, it has.

(She smiles. After a long look she turns and walks off.

Stephen...alone. He dances a little dance by himself. Happy.

He glances over to the wheelchair (which has been there the whole play.)

He exits, with a little jump in his step.)

CURTAIN

Lightning Source UK Ltd.
Milton Keynes UK
UKHW010638180722
406010UK00002B/594